A LAYPERSON'S GUIDE TO CREATION AND EVOLUTION

(Confessions of an Admitted Einstein Wannabe)

D. Vincent

Copyright D. Vincent, 11-10-18

No part of this book may be reprinted without written permission from the author and publisher.

Granite-Collen Communications
P.O. Box 621
Camarillo, CA 93011

INTRODUCTION

This work is based on a set of ideas, most of which have been well known and proven accurate for many years. My reason for writing this is not to restate, lay claim to, or challenge these ideas, but rather to propose a way in which they may fit together into a sequence of events which I find to be a perfectly logical theory that explains the creation of our universe and human evolution.

Other than a lifelong personal interest in science, space and astronomy, I have no qualifying background in these areas, which makes my credentials for writing this virtually non-existent. Why then even consider such a project? It's very simple. Being a writer, and a dedicated "Weekend Einstein Wannabe", I am personally fascinated by the amazing possibilities these ideas present, and I am driven to share them.

TABLE OF CONTENTS

A BRIEF HISTORY	5
THE THEORY	10
BACK IN THE DAY	14
A RECIPE FOR LIFE?	21
ANOTHER VERY LONG JOURNEY	32
RELIGION: A BRIDGE OF HOPE	42
SCIENCE AND TECHNOLOGY: THE FINAL STEPS?	52
A "SECOND LIFE" IN A CYBER-WORLD	54
FITTING THE PIECES TOGETHER	60
THE CRITICAL CAVEAT	67
NOW, ABOUT THAT PURPOSE:	69
EIGHT PREDICTIONS	72

A BRIEF HISTORY

One of the fondest memories I have of my father is a telescope-building project we shared when I was twelve. We bought two circular glass "slugs" and ground our own telescope mirrors. Using finer and finer grades of carborundum, over a period of weeks, the slugs slowly took on the slightly concave, highly polished shape needed to act as the light gathering, magnifying elements in our soon-to-be Newtonian reflector telescopes. When the grinding and polishing were done, we had the mirrors silvered, brought tubing, hardware, lenses, mounting equipment and put our scopes together. My equatorial mount was made from a length of two-inch galvanized pipe, a forty-five-degree elbow, and an old tire rim.

The many cold nights I gazed through that telescope in my backyard were some of the most exciting of my life. The craters on the Moon, the rings of Saturn, the moons of Jupiter. The Andromeda galaxy, star clusters, the Orion Nebula, shimmering clouds in the constellations of Scorpio and Sagittarius at the center of our own Milky Way galaxy. Today, at seventy-two, I still get intense feelings of wonder and excitement when I remember for the first time seeing these objects, dim, but no less magical, through the telescopes we had made. But that was the late 50's.

Fast Forward 60 Years...

Not long ago, I began to suspect what I eventually decided was a simple but amazing idea – a perfectly logical way everything (yes, literally *everything*) could have come into being.

It began to emerge one Sunday when my wife and I were working on a home improvement project at my son's house. I remember it was a sweltering summer afternoon and I was in his backyard, bent over stacking bricks, when suddenly, a simple but powerful idea occurred to me. I stopped, looked up at my son and said something like: "Hey, I just realized something. It's not a matter of if, but when."

"Huh?" he said with a quizzical look.

"If the Earth and the human race don't get destroyed, we humans *will* eventually discover the answers to those paradoxical questions about our origin, our purpose and what happens after we die! Maybe even in our lifetime."

He chuckled. He'd heard this kind of talk from me before. "After we die?" he said, with a condescending raised eyebrow.

"Right," I said.

"And how will we do that?"

"Our knowledge base is growing at a crazy pace, right?"

"Sure, I guess."

"So, it stands to reason that if we don't get wiped out by Artificial intelligence "units", some cosmic catastrophe, or incinerate the world with nuclear weapons, it may not be long before we'll reach a state of 'Escape Knowledge.' If we last long enough, it's inevitable. We'll have the true answers to our origin and spiritual questions."

He chuckled again. "'Escape Knowledge?'" And in the meantime, we just do nothing?"

"Right. We just continue to be curious, intensely driven human beings. Nature does the rest."

"Gotcha, Dad," he said, with an obviously patronizing smile.

Well, that was the first of many exchanges that went on for several months and, in the process, expanded into the ideas you're about to read. My wife will still have none of it (she's also heard more than her share of this Einstein Wannabe talk from me), and my son remains, to put it lightly, skeptical. None the less, I am about to tell you what I believe is the story of our origin…

THE THEORY

Is there a God? Were we created in human form or did we evolve? Will we ever discover the answers to these and the other paradoxical questions about our humanity? I believe the answer to these questions is yes, and I have a very simple idea that offers, if not proof, at least a darn convincing argument. It is based on a theory:

THEORY OF PURPOSEFUL
CREATION AND EVOLUTION

**Our universe came into being with a purpose:
To create, sustain and refine living Things.**

Does the word "purpose" in this theory imply some form of influence with the intent of seeing the "purpose" fulfilled? A completely random, undirected universe would have no purpose, right? It would simply come into being from natural, unplanned, unimagined events and exist just, well, "just because *it does!*" So, does a "purpose" suggest the involvement of a god or religious entity of some kind? Maybe, maybe not. To some degree that depends on your definition of the word "God". And that is, at least partially, what this exploration is all about – moving closer to a revised definition of that iconic word compatible with both religious and scientific beliefs.

Let's start with some basic scientific facts and what I consider a few reasonable assumptions.

Two Views: Micro and Macro

In their search for the answers to questions about our origin, scientists often explore the tiniest bits and pieces of our universe – the atoms we are made of, the particles those atoms are made of; the particles those *particles,* or strings or waves, are made of. In other words, the nearly infinitesimally small states of matter (and anti- or dark matter). From these sub-microscopic, unimaginably tiny explorations, scientists have calculated, theorized and discovered verifiable facts and offered profound, theories about the origin and functions of the universe. I tend to think of this way of exploring our universe as a micro view.

What you will find on these pages is more of a macro view, a discussion of, for the most part, the very *large* things in our universe and a general notion of how I believe they have come to be and interacted, driven by that purpose I mentioned.

I do realize that formal scientific research, in whatever form it takes, is carried out by men and women with extraordinary minds, sophisticated equipment and years of education and experience. As I mentioned, this is not an attempt to contradict what these professionals have discovered. What you will read here is simply my idea of a universal *umbrella design* – a very broad description of a model, under which most, if not all of the current scientific facts and theories remain intact. In this design, particles still collide, pop in and out of existence, and move in steams and/or waves or both. Schrödinger's cat remains alive and dead at the same time in its paradoxical box. The detailed Relativity, Special Relativity and bizarre Quantum Mechanics laws that physicists have developed in their attempts to define our universe, with all their mind-bending paradoxes, remain as they are.

I am simply theorizing that all of those things together make up the *ingredients* in a self-perpetuating cosmic recipe that does, and will continue to do, three things spontaneously and inevitably as long as it remains in existence:

Create, sustain and refine living things.

BACK IN THE DAY

At some point in the very, very, *very* distant past, in fact, so far in the past it was before our universe existed, an "explosive event" of unimaginable speed, size and power took place (scientists refer to it as a rapid expansion, since at that point there was nothing to explode).

Most scientists agree this is how our universe began, and they have named that expansion the Big Bang. The Big Bang originated from what these scientists call a Singularity. I will not pretend to be smart or educated enough to explain exactly what a Singularity is, however, I do know that it is described as an incredibly dense and, in the case of our Big Bang, almost infinitesimally tiny point. A Singularity is also another name associated with Black Holes, by the way, which are incredibly dense sort of bottomless pits in the fabric of the universe, pulling in everything nearby with such enormous gravitational power that not even light can escape it.

A Tale of Two Bangs

I think we can safely assume there were at least two versions of how that Big Bang _could_ have taken place. I call these The Straight Bang and The Curved Bang.

The Straight Bang

Imagine the Straight Bang as a tremendous explosive expansion that blew enormous globs of blazing universal material straight out into the nothingness of a newly created space. And because in this scenario space truly was nothingness, all those cosmic chunks and blobs and pre-atomic goop blew *straight away* from the center and kept going in straight lines at incredible speeds. You could imagine this as something like the initial blast of one of those huge, colorful, 4th of July fireworks explosions that at first sends fiery points of light straight out in what appears to be a perfect circle.

The Curved Bang

Now picture a second version of that expansion. It begins the same, but in this case all that cosmic stuff *does not go straight a*way from the center. That's because in this scenario space is not nothingness. In fact, space, according to Albert Einstein (I think we can trust his definition), is actually Space-Time. And two characteristics of Space-Time, very loosely defined, are first, that it is something like an immense, curved fabric stretched tightly across the entire expanse of the Universe. I have to qualify that statement and say this "fabric" is not exactly like a single, two-dimensional sheet as you might be thinking. It is actually much more complicated than that, but again, for purposes of this story, if you imagine an enormous taut sheet of fabric, that will do just fine. And second, this taught sheet has the capability to exert gravitational force.

The Crucial Difference

Now here is the important distinction between these two possible versions of that initial expansion. As this second version (the Curved Bang) takes place, the chunks of stuff shot away from the singularity are so huge and dense that they ripple and warp this curved, universal, gravitational fabric, and that causes the blazing hot materials to arc and swirl and *curve* in those ripples and warped areas. You might picture this version of the expansion as something like a tremendous blast of hot universal blobs shot away from a center point and deflected in the ripples of surrounding, invisible sheets.

And because these arcs and curves (instead of straight lines) were created in the very beginning, they set the stage for life to follow. In other words, we could think of that one simple condition and action, *the motion of a curve,* as the representation of a cosmic seed taking hold in the womb of a brand-new universe. And it was from that seed (though indirectly as you will see) that all life as we know it has sprung. How?

Orbits!

As I mentioned, these large pieces of stuff slung out into space were very dense. And it is the density (the mass) of objects that creates those pits and ripples and indentions in the gravitational fabric, which, as you may have guessed, are areas of gravitational influence.

Picture a bowling ball sitting in the center of a tarp that has been stretched tightly across something like the opening of a large round hole. The ball's weight and density creates an indention in the fabric. If you roll a marble sideways out onto this tarp, its path will bend. It will be drawn toward the bowling ball and begin to circle in toward the indention, suggesting an orbit. That's essentially how gravity works in space, and it is why planets orbit suns and why moons orbit planets, our moon included.

But had all that material blown straight away into nothingness as in the Straight Bang scenario, planetary orbits could not have formed. And without orbits, the conditions that would later allow life as we know it to begin could not have emerged. The most important of these conditions, by the way, was the formation of what scientists refer to as "Habitual Zone" obits – warm, comfortable areas for orbiting planets, just the right distances from their suns – which allowed liquid water to form and with it the precursors of life.

The bottom line? Without the essential curves, the stuff of the universe would still be going in straight lines today, and we would not be having this conversation!

With that in mind, let me offer a foundation for my theory:

The Big Bang: A tremendous, explosive expansion created and sent massive amounts of dense universal material out into the curved gravitational fabric of our newly created, Space-Time Universe.

Chunks and blobs and bits and pieces of this dense, unimaginably hot cosmic material swirled, rippled, arced and curved in the fabric.

This "stuff" of the universe slowly began to cool, condensing, coalescing and solidifying into planetary objects that began to circle in around suns, until…eventually, in addition to galaxies, nebula, clusters, black holes and many other cosmic objects, *planets formed* – some of which settled in what scientists refer to as life-friendly "Habitable Zones." And we are just now beginning to realize that it's virtually *countless* planets – orbiting at a

minimum, trillions of suns throughout the universe.

I think most scientists would agree this is generally how we believe the Big Bang took place and these events are generally assumed to be what followed.

A RECIPE FOR LIFE?

Included in these cosmic materials that were swirling, forming, coalescing, cooling and orbiting, were all the basic atomic and molecular ingredients of life as we know it. I don't have a complete list, nor do I have details of exactly how and when they formed. But again, I believe those are facts we can do without. Here's the important fact to keep in mind:

> **Just like the solar and planetary materials settling into their orbits, all these atomic and molecular ingredients of life were also being created and spewn into space following the Big Bang.**

Carbon, liquid water and oxygen were a few of these many key ingredients, and scientists believe that when conditions were just right on planets in Habitable Zone orbits, a spontaneous molecular process began to take place. These stuff-of-life-molecules began to coalesce and assemble into what became the emergence of life as we know it.

Now, here is a critical question:

> **Did all this come about by *chance*? The curved, universal gravitational fabric, the essential curved trajectories leading to the formation of stars which provided light and warmth and life-friendly habitable zones for planets circling around them? The molecules that began to coalesce and assemble? In other words, all the ingredients, materials, conditions and locations needed for life to begin? Coincidental?**

As you will soon learn, I am not a religious believer. I *do* believe in Darwin's theory of evolution. I *do not* believe that a god, as we typically imagine "Him," created man, woman and the universe. However, I find it extremely unlikely that the sequence we're beginning to explore could have come about without *some form of initial influence.* In my mind that would be like saying, for no reason and with no direction, numerous exotic ingredients suddenly popped into existence in your kitchen and combined by themselves in just the right amounts, under just the right temperature conditions, at just the right place (inside your oven, which just happened to be pre-set to an ideal cooking temperature), for just the right amount of time, and emerged as a delicious casserole! I don't think so. You will form your own opinions, of course, but we're just getting started on this exploration. Let's follow the story a little farther down the evolutionary road and see where it leads us.

The Probability of Life

As I've noted, most scientists and astrophysicists believe that the emergence of life and the process of evolution were completely random events. But how can that be?

How can a chaotic, unsupervised jumble of cosmic rocks and fireballs swirling, banging and colliding and exploding in space possibly lead, without the help of some outside influence, to intelligent, self-aware creatures like you and me? Interestingly enough I believe it is, at least *in part,* a simple matter of numbers! Really, really, ***really*** big numbers!

Space and Stuff– Lots of It!

Try to imagine the size of our entire universe. Our Galaxy alone, The Milky Way, is roughly 100,000 light years across. A light year is the distance light travels in one year. Light moves just a bit over 186,000 miles per second. Multiply this by 60 to arrive and the distance it travels in *one minute–* 11,160,000 miles. Now multiply that number times 60 to find the distance in one hour– 669,000,000miles. Then multiple that number times 24 hours to get the distance light travels in one day– 160,704,000,000 miles. Then times 7 for a week, times 4 for a month, 12 for a year, and finally times 100,000 of those years for that beam of light to take an end-to-end ride across our galaxy. And though working with numbers like that would set our iPhone calculators to smoking, they are literally *a speck* in the total space of our universe – which scientists say contains over a trillion galaxies like our Milky Way. In other words, distances so immense and areas so vast that we can't possibly grasp their enormity; virtually countless suns out in these unimaginably vast expanses, and circling around them groups of those orbiting planets I've been describing. As you can see, the numbers become truly mind-boggling!

Which, leads to another aspect of this theory:

> **The laws of probability suggest that because there are virtually *countless* planets, and planet-like objects out there swirling around in the stuff of life, in a space so vast it is unimaginable, there is a very high probability that, although in most cases the emergence of**

life will fail (we'll talk about why shortly) we can virtually be sure that everything will happen just right on at least a relative "few" of those planets allowing life to emerge and begin to evolve.

Now add to this thought…

"Countless" Years.

Right – those countless planets in that immense space, have been cooling, orbiting and mixing with the stuff of life for virtually *countless years* – in fact, billions upon billions of years! And what effect does this incomprehensible amount of time have on the probability that life would emerge out of an immense, undirected universe, churning with cosmic chaos, but having plenty of fertile materials?

A little detour will complete the idea.

Getting It Right on the High Seas!

Imagine a small, specially crafted boat afloat on rolling seas. It has a perfectly flat, smooth deck that is fifty feet long by twenty feet wide. On this deck are a thousand colored marbles: blue, green, yellow, brown, black, turquoise, pink, and many other colors. In addition, there is a single red marble. Let's assume the colored marbles represent the planets and the ingredients of life swirling around in the universe. Let's also assume the red marble is extraordinary. Under certain conditions it has miraculous, life-producing qualities, and thus it may represent the key to what might be the emergence of first life. I say "may represent" because the red marble has another interesting characteristic. It can only initiate the emergence of life if it touches 5 colored marbles in a specific sequence and with perfect timing – blue, green, yellow, brown and black. Touching these marbles in this order for a split-second each – and *only* in this way – turns on the red marble's life creating properties, allowing it to go into action.

The boat is tossing on the rolling waves and because the surface of the deck tilts continually in virtually all directions, marbles are rolling around randomly with no possibility of order or control over their paths. And there is one final aspect to this scenario. The boat will only be afloat on this sea for *one hour*.

Now, imagine that you have a bird's-eye-view of this deck. What would you say are the chances are that at some point you would see the red marble hit each of the prerequisite 5 colored marbles in the correct order and timing and begin to create life during this short period of time? Pretty slim? In fact, that's an understatement. Why? Because there is a very short amount of time and the number of colored marbles is not, relatively speaking, very large, it would be highly improbable that this sequence of hits would take place. And that means the probability that life would emerge on this boat is virtually non-existent.

Supersized!

Now imagine another, much larger boat. In fact, this one is a tremendous barge afloat on the same rolling seas. Like the smaller boat, the deck is a flat, smooth surface, but in this case it is tremendous, roughly the size of 100 city blocks. On this immense deck are a 100 billion colored marbles and a 10 million red marbles. And in this case, instead of an hour, the barge will be afloat for a *billion years*. Again, you have a bird's eye view. Would you say that on this ship the probability that during that billion-year time frame you would see one of the million red marbles strike 5 specified marbles in the correct order and timing? If you said yes, you're right. Because of the astronomical number of years and the virtually countless number of red and colored marbles rolling around on this deck, even though the entire scenario is *undirected* and *completely random* (thus seemingly without purpose), the probability that at some point a red marble will strike the specified 5 marbles in the correct order and timing and life will emerge increases dramatically.

If we use these sea-going examples as analogies for our universe following the Big Bang, we can add the following to my theory:

> **Although the Big Bang process was chaotic and completely undirected, because there were *so many possible planets* on which life could emerge, *so much fertile material* forming in the immense universe, and the *length of time* they all interacted was *so incredibly long*, it was inevitable that at some place, or places in the universe conditions would combine just right, life would emerge and evolution would begin.**

On the vast majority of those planets the specific conditions required for the creation and evolution of life as we know it have not materialized and never will. That's because many are too close to their suns so they are too hot, and many others are so far out they are too cold. Still others are what scientists call gas giants with thick, toxic atmospheres hiding deep "gooey" cores.

There are also large and small asteroids, dwarf planets, comets, unstable and elongated orbits, places where the essential ingredients required for the beginning of life as we know it are simply not present.

Evolution?

But what exactly is this process of evolution we just touched on that allows (as we will see) only a few of many life forms to survive and evolve.

ANOTHER VERY LONG JOURNEY

Once life does begin, it appears the process of evolving from simple to more complex life forms also began spontaneously. And this raises another interesting question: Did *spontaneous evolution* also happen by chance? Once life begins (or *as* it begins), does a functional system of sustaining and refining all living things emerge, again, for no reason and with no purpose? Or did this functional system come about with a *purpose – to nurture life and keep it advancing*? And if so, was it part of a much broader "recipe?" I'd say yes. But whatever the answer to this question, there is one thing we can all agree on (excluding those religious believers who deny that any of this took place): *Evolution adds even more astronomical amounts of time to the process. It is an extremely long, slow and violent process.* It is fraught with merciless brutality and constant life or death struggles. In our case, it has taken about 3.5 billion years, and most (but not all) scientists agree, it has progressed by means of the bloody process of change Charles Darwin called Natural Selection. The human evolutionary line has evolved through and survived this process (to this point, at least) but it's a good bet that most life forms that have sprung up here on Earth and out there on those other likely planets have not.

Why? Consider another aspect of the theory:

> **Over *so many years*, (yes, there's that extraordinary time factor again) natural events like large meteor strikes, massive volcanic eruptions, exploding suns, escaping atmospheres and many, many, *many* other types of cosmic catastrophes, not to mention continual battles for species survival, would almost certainly, as a natural result of cosmic existence, destroy most fledgling planetary life forms before they got too far down the evolutionary road.**

One of many examples right here on Earth is the dinosaurs. Most scientists believe a large meteor impact in what we now call the Yucatan Peninsula changed the temperature and weather patterns on Earth so drastically that the dinosaurs (and many other life forms) could not survive. In fact, scientists have discovered that about 99 percent of *all* species that have existed on Earth have gone extinct because of catastrophic events or the inability to compete effectively and pass on their genes.

None the less, our universe seems to be designed to keep trying, time after time after time after time, endlessly repeating that same life producing process wherever the conditions are right. The result is that today billions of life forms exist on Earth. Just how, you ask, have *they* managed to survive?

The Sea Turtle Parallel

As we all know, a process of natural balance exists here on Earth – one that I believe is also a universal model for the continuation of life. Here is one of the many *earthly* examples.

Every year, thousands of newly hatched sea turtles break free from their eggs in the sands of tropical beaches. Many immediately become the victims of crabs, fish, powerful currents, weather and other dangers, and as a result, die shortly after their birth. Others last through the initial stages of life and perish in various ways – often as the victims of voracious predators. Only a relative few make it to maturity.

Through this process, nature limits their numbers. In addition, those that die become sustenance for other aquatic travelers in the oceans. In this way, a natural, symbiotic balance is maintained that allows the turtle species (and many other species) to continue to exist and evolve while remaining at sustainable levels. It also allows the oceans to remain a rich, open environment for all aquatic creatures. By contrast, if *all* baby turtles grew to maturity, mated and gave birth, it wouldn't take long before they would overrun and destroy the oceans.

Now, let's assume I was correct when I suggested earlier that most of those new life forms created on orbiting planets out in the universe would die during their early evolution from natural events and competition. We can illustrate how these ideas parallel our turtle story:

Because our oceans are extremely open and vast...

(just as the universe is extremely open and vast)

...and a relatively small number of mature sea turtles move though its vast waters...

(just as a relatively limited number of life's recipe elements exist in the immense universe)

…the number of times mature female turtles mate and become fertilized is limited…

(just as the number of times the recipe elements merge on an orbiting "Habitable Zone" planet, everything happens just right and life spontaneously emerges).

Later, after those female turtles have laid their eggs and hundreds of baby turtles begin a new cycle, hatching in the sand on those tropical islands as they attempt to reach the sea most, as we have noted, become victims and others die at various stages of growth. The result? Again, only a relatively small number of sea turtles live to maturity…

(just as various cosmic events limit the process of evolution in the universe, wiping out most

"infant" or growing life forms, leaving relatively few to reach maturity, thus allowing the cycle to continue).

Assuming this analogy is valid, it leads to yet another interesting question: Could those planetary life forms that perish be providing some form of balance and sustenance for other newly emerging life forms? As an example, had the dinosaurs *not* emerged on earth then gone extinct, and as a result we humans had *not* discovered and been able to use fósil fuels to power our progress, would we have been able to progress as we have? Could the remains of the dinosaurs and early plant life be a random form of sustenance that in our case helped further the advancement of our evolution?

Natural Selection: The "Endless" Gauntlet

According to Charles Darwin's theory of Natural Selection (the most agreed upon theory by scientists, though it does leave some unanswered questions), the fittest generations of a species benefit from positive genetic mutations which allow them to evolve and compete more effectively in battles for survival. As one example, the most aggressive and successful male lions on the African Savannas will often be genetically superior to other males, and they will most typically get the best mates. Their cubs will come into the world with superior genes. As they grow up and do the same, and this happens repeatedly over millions of years, the process of evolution produces very slow changes – in some cases advantageous gene mutations that improve the lions' ability to compete and thus stay successful. However, the mutation of genes is not always positive for the species. Sometimes it is harmful and those lions who inherit the faulty gene mutations will not fare as well in the survival category. Bottom line? The creatures with positive genetic makeups tend to thrive and slowly improve and advance their genes and thus the species.

Competition – The Brutality Factor

The basic driver of this process of natural selection is *competition* – simply put, the competitors fight it out and the winners (often those genetically superior) get the prizes – the mates, the food, territory, and of course, the opportunity to pass on those all-important superior genes.

Most scientists agree that as billions of years have passed since life appeared on Earth, living organisms have been competing in life and death struggles and experiencing both positive and negative genetic mutations. The positive ones, we believe, have helped lead to successful, increasingly complex life forms, with one culminating 3.5 billion years later as us human beings! And in our case alone, after nearly 4 billion years, we are the only living things we know of in existence that have reached a level of self-awareness. That means at some point in our evolution, tens (or perhaps hundreds) of thousands of years ago, our primitive ancestors had become intelligent enough to the realize that they were living beings under a mysterious, blinding ball of daylight, gleaming points in the night sky, and a silver ball that came in the dark, changing its face and position repeatedly as the light and darkness cycles passed. In other words, they were becoming *self-aware* of an unfathomable mystery.

RELIGION – A BRIDGE OF HOPE

Imagine what it must have been like for the first primitive humans who began to realize that! Who were they? What were they? Where had they come from? Who or what had created them? And perhaps most important: *What happened when they died?* It makes sense that these questions would have been terrifying and overwhelming to our early ancestors (and, in fact, still are in many cases today). It also makes sense that religious myths and beliefs would emerge at this phase in our evolution, since they could provide the answers to those frightening questions of origin, purpose and mortality. A belief in "The Gods" resolved these questions for a still relatively ignorant (I don't mean that word in a negative sense), adolescent race. It also conveniently placed us humans at the center of the universe! What more could we have asked for in terms of comfort, security and a way to alleviate the frightening possibility that we were alone in space? An all-powerful parent-guardian (a mythological god) to love and guide us – and eternal life-after-death thrown in to boot!

It was just what we needed at the time, and boy did we believe! Gods of the sky, gods of the earth and sea. Gods of war and the moon. Gods of love. Happy gods and angry gods causing famines and storms and gods requiring animal and human sacrifices! Eventually, single gods, including Christ, Buddha and Allah, became the icons of established religions and symbols of the sacred truths. Though these icons remain absent and shrouded in mystery to this day, we have believed ever since those first revelations that these gods have

"The Answers," and if we abide by their rules, if we follow their commandments, we imperfect "children" will be given those answers someday in a glorious, endless afterlife. Sounds fantastic, right? That's why it has been our spiritual cornerstone for many thousands of years.

Intellectual Awareness – The Religious Spoiler

But "lately" things have changed. As the centuries have passed, we've learned an enormous amount. As a race we have matured. We are no longer restricted by the "innocent ignorance" of our adolescence, and our intellectual momentum has eroded those religious myths and beliefs. These days, proven scientific facts tell us that the traditional religious answers, those known only by the gods and kept hidden until our afterlives, do not exist after all. They were a form of rationalization and support appearing at a critical time in our evolution. And they have done their job wonderfully, sustaining us and providing a bridge over the dark uncertainties of our emerging self-awareness, until we reached (in fact, may just now be reaching) the next phase in our evolutionary journey.

Today our heightened intellectual awareness has led to significant discoveries, and these have compounded in recent years, thus broadening our knowledge on an exponential scale. In short, we have "suddenly" (over the past several hundred years) begun to understand DNA, genetic codes, diseases, the nature of Earth and objects in space, time, motion, physics, relativity, atomic and subatomic particles, quantum mechanics, laws of nature, evolution, and so on, and this immense wave of understanding continues to broaden our knowledge *at an ever accelerating pace.* And with all this new knowledge under our belts, we are fitting the facts together into what most scientists consider a much more accurate model of our origin and future path.

Because of this new science-based model, my guess is that traditional religions will continue to weaken and slowly become obsolete. This will not be a quick or easy transition. In fact, it may be bloody and violent because we've practiced religious traditions for thousands of years and they are heavily ingrained in the human psyche – but the transition *is* happening, and it *will* continue. One way to possibly limit or eliminate the imminent violence, may be to somehow re-define the word "God," forming a definition that will satisfy both the scientists and religious believers. My hope is that in some small way, this essay might help support and further that idea

Competition – Like Religion?

Earlier, we touched on the subject of competition. And you'll recall I noted that it was the basis of the brutal process of Natural Selection and Evolution. But as we have gained knowledge and become intellectually aware of our world and our nature, we've begun to realize that although competition, just like the gods and traditional religious beliefs, has served a very important role in our evolution, it, too, is becoming obsolete for us humans. Here's why.

Competition is nature's evolutionary engine, driving and sustaining an *undirected, openly evolving environment*, in which *wild organisms* battle for food, mates (most importantly the ability to pass on those "best" genes), survival, shelter, and so on. As I've noted, this is how *species have evolved.*
But living in *controlled* communal societies, and becoming continually *less wild* over many millions of years, *we humans have been increasingly regulating and limiting the natural process of competition.* For instance, we take care of our weak and defenseless. We allow them to live longer, more rewarding lives. This is a form of empathy and compassion that doesn't fit the brutal process of competition. We learn to respect and value *all* life, including animals, which is also not consistent with the competitive process. We domesticate

animals, farm for our food, and feed the masses. These and many other types of modifications to the natural, "wild" process of competition for survival, have been broadening for thousands of years, and they are becoming continually more prevalent and ingrained in our way of life.

Revealing Trends

Here are a few modern trends that illustrate this idea: Think about males and females in today's global societies. Just as competition is slowly weakening, aren't the traditional male and female roles –which, remember, *evolved from our ancestral animal origins, that happen to be a major foundational element of the competitive process* – slowly becoming intersexual or androgynous? A few examples are the growing numbers of gay, lesbian, bi-sexual, fluid and transgendered individuals being accepted in societies all over the world. Are the less prominent male and female roles shrinking because the need for competition is waning?

Also, females, once extremely limited by the rules of society, competition and religion, are moving into more powerful, decisive, traditionally "manly" roles (although, admittedly, some societies are far behind on this). Ask yourself these questions:

> **Aren't males more suited to be leaders in highly physical, confrontational, combative environments – which describes our distant and recent past? And...**

> **...aren't females more likely to practice leadership through non-violence and empathy, thus nurturing peaceful cohabitation – which is what we currently see emerging and the direction our societies will likely take as they evolve?**

I realize these trends reflect only a microscopic flash of time in the immense timeline of our evolution, but because they fit perfectly with the logic and progression of much larger changes, I'd be willing to bet that as time goes on we will see more and more female leaders (Ladies, it appears you are next up, and that will be a good thing) and intersexual individuals will also increase as the years roll on.

Remember, too, the number of marriages is continually declining. So, what about the traditional family unit that follows courtship, marriage and mating? For thousands of years it has been a central, foundational element of our societies, but just as sexual roles have begun to blur, isn't the family unit – *which is based largely on those traditional male-female role-models and thus competition*– becoming increasingly less prevalent in our society? Isn't the traditional family unit slowly disappearing? Though many people, particularly religious conservatives, would tell us that an intersexual society and the loss of the family unit are proof of societal decay, my opinion is different. I believe these developments, though difficult to accept and no doubt problematic in their early stages, are signs that we are moving in an *inevitable,* and hopefully more beneficial direction. You'll understand why shortly.

But what about reproduction, you ask? If it's true that we're becoming more androgynous, how will we eventually pass on those best genes? Well, at this point we can have all the sex we want for sheer pleasure (in whatever form it takes), and in spite of our increasing "uniformity," we can conceive offspring (as we do today) with harvested sperm and eggs.

My guess is that eventually it will become clear to us that the competitive process, including its various facets – like male-female roles and the family unit – is one we are phasing out of simply because we re learning that *we can control the path of our evolution without the bloody, merciless process of competition in Natural Selection.* And if you look around at today's world, you will realize that the two primary ways we're applying that control are through science and technology.

SCIENCE AND TECHNOLOGY:
THE FINAL STEPS?

Technology, as we know, is progressing at an incredible pace, accelerating almost daily. We are now able to communicate with our hand-held devices at the touch of a graphic screen. Human thoughts have been transmitted wirelessly across long distances. We are creating more and more "body parts" and implants. These days we have robots that serve as butlers and doctors and tiny cameras in capsules that move through our bodies "filming" our digestive systems. We also have cars that can drive themselves. We've produced clones, human/animal hybrids and mapped the human genome. The products of computer technology are rapidly shrinking in size, while at the same time becoming more sophisticated and powerful, and (this is a critical one):

Cyberspace is becoming, not just a trend, but a seamless, integrated part of our *existence and evolution*, especially for the younger, emerging generation.

Meanwhile, we are depending less on our bodies for survival. Obesity is increasing; partly because we *move much less* than we used to, and my guess is that this trend of increasing physical sedation and deterioration will continue (in spite of all those New Year gym memberships). The good news is scientific breakthroughs are now allowing us to repair, transplant and create body parts. Synthetic tissues are being produced in labs as I write this. It appears diseases like Cancer, Alzheimer's and Aids will soon be cured or fully controlled. Life will be sustainable for much longer periods of time – possibly exponentially. Computers and other technological advancements will continue to become increasingly prevalent, concentrated and integrated into our lives.

And where is all this leading us? Consider a very possible and profound eventuality:

Humans or Machines?

In the near future, the line between natural humanity and manufactured life will become blurred. I recently had two shoulder replacement surgeries. Both of my shoulder joints now consist of titanium balls and plastic sockets. That's not much in relation to my overall body mass – maybe 1 percent? But at what point is a person still a natural human being if, say, 50 percent of man-made elements make up his or her body – bones, joints, skin, organs, muscles tendons, etc.? How about 70 percent? 90 percent? Suppose it reaches 95 percent? Suppose at some point the only non-manmade element inside a human being is the brain? Is he or she still a "human" being? Or, perhaps a "humanly-*made*" being? Or an android with a human brain? And this begs yet another question: What exactly *is* "humanity"? Body *and*

mind? Only mind? Suppose something equivalent to the human *brain* is eventually
be produced? Artificial Intelligence is making incredible advancements, and even today computers are capable of learning and reasoning. What *then* would humanity be? A digital, synaptic consciousness of some sort? Would we even *need* our troublesome, continually aging bodies in this advanced technologically-driven future – be they robotic, man-made or natural – since we are finding less and less use for them?

A "SECOND LIFE" IN A CYBER-WORLD?

Wikipedia has described the online game known as "Second Life" this way:

> "*Second Life is an online <u>virtual world</u>, developed by <u>Linden Lab</u>, launched on June 23, 2003. The users in Second Life, called <u>Residents</u>, can interact with each other through <u>avatars</u>. Residents can explore the world (known as the grid), meet other residents, socialize, participate in individual and group activities, and create and trade <u>virtual property</u> and services with one another…*"

> "*Built into the software is a <u>three-dimensional modeling</u> tool based on simple geometric shapes that allows residents to build virtual objects. There is also a <u>procedural</u> scripting language,*

> <u>Linden Scripting Language</u>, *which can be used to add interactivity to objects.* <u>Sculpted prims</u> *(sculpties),* <u>mesh</u>*, textures for clothing or other objects, animations, and gestures can be created using external {and imported} software.*"

Right. A computer video game.

Now imagine a futuristic, highly advanced version *not of this game,* but of *a virtual reality experience that offers vivid and total immersion achieved through technology.* If you decide to "play" you experience a complete disconnect from life as we normally live it and move into in a cyber world just as "real" as what we now experience. If that sounds like something you'd like to experience (remember how avid our younger generation is with computer "games"), before you embark on your personal cyber journey, stop and think about the ramifications this presents.

If the world you enter can become a kind of sanctuary of your own design, a "real" place you've actually been able to create and construct exactly to your liking, if you meet others like yourself and interact in this "other life," and the often difficult, psychological baggage of "human life" doesn't exist, wouldn't you want to stay awhile? Wander this incredible world? Check things out? Return often? Might we as a race, begin to migrate *over a long period of time* into such virtual worlds? Would we decide to enter and stay for longer and longer periods of time, creating alternate lives?

And could more advanced technologies evolve and lead to freeing us from our problematic, high maintenance bodies? Could we eventually find a way to release the essence of what we are – even *without* the fleshy gray matter of our brains and the role-play-like software experience? Could the "game" *itself* evolve, melding with our consciousness?

If you are chuckling and shaking your head at this point, I can't blame you. But stop and think about your consciousness *right now*. You perceive and experience your world through the senses and consciousness created by your mind – an extremely sophisticated "flesh computer," How much different would it be if our perceptions were provided by *non*-flesh, but *equally* (or perhaps more) *sophisticated flesh-<u>like</u> computers*?

Too far out? An unbelievable leap of the imagination? Maybe. But if you're willing to take that leap – if you can let your imagination "out of the box" and envision what incredible advancements we've evolved through even up to this point, along with those that are likely on the near horizon, a progression something like this sounds much less bizarre.

FITTING THE PIECES TOGETHER

As I've noted, I'll be the first to admit these ideas are unproven stretches of the imagination and I'm just a guy (I'm sure not the first) who's pitching them out. But what draws me to these conclusions? When I allow my mind to "travel" the immense, full sequence from the Big Bang through subsequent evolutionary events, along with historic and current patterns and trends I see happening around me in life, they form a perfectly logical and, I believe, highly likely, sequence of events encapsulating the creation and sustenance of life. At the risk of a little more repetition, here it is one final time:

> **A tremendous explosive expansion produces a random, ungoverned, "free-functioning" model designed for one reason and with a single purpose -- to create and refine**

virtually countless living organisms throughout the new Universe.

Though the vast majority of those organisms fail and disappear, given countless planets, and taking place over the course of billions of years, a "few" evolve through into sophisticated forms. And this universal, undirected model, eventually, spontaneously produces a species that evolves to a state of self-awareness.

That self-awareness launches us humans on an *irreversible pursuit* to gain more and more knowledge and eventually discover our origin, purpose and the answers to the questions of existence.

If there is a chance all this might be true, isn't it *extremely* un*likely* that this multi-faceted, *purpose-driven* universal model would come about simply by chance?

One big question this raises, of course, is: If not by chance and not for a purpose, then *why?*

Why would a sequence of events so immense and perfectly un-choreographed, constantly produce and refine living things? Events that seem to be preparing and leading those beings to experience some future change? For no reason? Just because the universe exists?

As I've said, I am not religious. But if not by chance and not by a God, then how and why…?

> ***If there is no god controlling us or the process we've been discussing (which is my belief), could our Big Bang <u>itself</u> have been the single "designed" event– an intended, predetermined, big bang sequence in which all the elements required for life and evolution were condensed and combined into a kind of immense, self-functioning cosmic recipe <u>designed</u> to create and refine life spontaneously, given enough***

> ***time and materials, with no external control or direction?***

We have chemical recipes for making everything from glue to perfume. Most common for many of us are food recipes which can be prepared virtually any place on earth at any time and achieve the *same result*. Where, when and how these recipes first appeared we may not know, but we know that if cooks in, say, ten different countries on earth combine the same ingredients (flour, yeast, eggs, etc.) under the same conditions the result will always be bread. In such cases, the process and end result, though spontaneous are common denominators wherever, whenever and regardless of how many times the recipe may be used.

The creators are not actually creators at all, but *users* or possibly *executors* or *distributors* of a recipe in a secondary sense. They are not necessarily a He or She or for that matter anything living. The "creators" may simply be "cosmic starters" of some sort – entities that did not directly guide or shape the individual elements and processes during the creation of the universe and our evolutionary life-cycle, but simply set in motion a predictable sequence (possibly unknowingly), which combined and processed specific ingredients under certain conditions and thus alleviated (again, given enough time and materials) the need for direction and guidance.

As we've noted, one key attribute that makes this recipe so amazingly simple, is that even though it is a completely undirected, random process, with:

> **Virtually "endless" amounts of time (billions of years) "countless" opportunities for life to begin (trillions of planets) and an "endless" supply of essential atomic and molecular materials all swirling together in space, the probability that the recipe will continually produce, sustain and refine intelligent living things, seems virtually assured.**

Where and when the recipe came into being who knows. And then there's the other big one: *Why?*

***Why* the intention to create, sustain and refine life?**

That, once again, is the crucial question and the one to which we seek "The Answers." As I mentioned in the beginning of this writing, provided we survive, given enough time, logic tells us we will eventually discover those Answers. However, that undirected giant meteor or comet could appear out of nowhere any time and wipe us out completely – which means we would become another one of those failed attempts at life in the Universe – a cosmic statistic, so to speak. But since the Universe keeps repeating the process regardless of the number of failures and close calls, I'd bet some other species or race in another corner of the universe will get farther than us and perhaps reach a finish line: The discovery of "The Answers."

Possible? Plausible? A-1 Kook Status confirmed? You decide. But as I've said, in my mind there seems to be no other conclusion. Logic tells us that continual, uninterrupted refinement of a species without extinction, *has to* eventually lead to some form of "perfection."

THE CRITICAL CAVEAT

There is one potentially tragic, but very possible variation, of this theory of life forms being destroyed by natural events. *We might just destroy ourselves!* Most people who lived through the nuclear build-up during the Cold War of the fifties and sixties would likely agree we came uncomfortably close to possibly wiping humanity off the face of the earth. Today, the possibility of a global nuclear wars is (let's hope) history. Instead, most scientists worry more about CO_2 emissions and the Green House Effect causing Global Warming. And they tell us that if we do not stop polluting our atmosphere, we will pay a heavy price – perhaps the *ultimate* price.

Global Warming, Politics and...the End?

I remember watching Discovery Channel's wonderful series, "Cosmos." In a program entitled "The World Set

Free," the host, Neil deGrasse Tyson, began with an image of the sweltering cloud-covered surface of Venus. He explained that scientists believe the planet may have been very much like Earth in its distant past – until something catastrophic happened. The Greenhouse Effect, triggered by Global Warming, turned Venus into an uninhabitable hothouse, literally "boiling" in its own lethal gasses. Global Warming? The same Global Warming we are currently arguing about in Washington? That's the one....

Whatever the future holds, one thing is for sure. Given how far we've come, we may just be on the verge of achieving our ultimate goal, learning the answers to all the paradoxical questions about our origin. And it would be an unimaginably shameful waste of nearly 4 billion years of evolution and genetic refinement if we do ourselves in now.

NOW, ABOUT THAT PURPOSE

At the outset of this essay, I proposed a theory:

Our universe has a purpose:
To create, sustain and refine living things.

I suspect many people still take issue with that statement – especially the second part. We'll get to that shortly, but let's start with a few questions, starting with the first part of the theory:

Our Universe has a purpose:

Everything we know of in our world has purpose. The seas provide water, rain, the stuff of storms, food, ecosystems and weather patterns; the land provides food and sustenance, animals provide food and keep a healthy natural balance intact; plants provide food and oxygen, the sun provides warmth and energy, and the list goes on. And how about us? Do we have a purpose? In a strictly utilitarian sense, it seems we have two purposes: To survive as a species and continue to reproduce and advance. Is it the same throughout the Universe? Does everything out there have a purpose? I believe so. But it's the second part of the statement that probably elicits the most skepticism:

To create, sustain and refine living things.

If you think about what is happening in the Universe, you may come to the conclusion (as I did) that the only thing it produces of consequence or meaning is life. The rest, as fascinating and complex as it may be for science, is, to the best of our knowledge, an immense, mind-boggling fireworks show – swirling, burning, warping, exploding and dazzling. Do we know of another purpose for the stars to form and planetary objects to orbit around them other than to sustain life? Do we know of another purpose for the ingredients of life – water, carbon, oxygen and many others – to combine and coalesce other than to create life? Do we know of another purpose for stars to explode in supernovas and send the stuff of life spewing out into the Universe? Do we know of another purpose for the Universe to be interwoven with a space-time gravitational fabric that makes orbits, galaxies and solar systems possible?

We can ask these questions about any and everything that happens in the Universe, from the smallest and most complex actions, to the largest and simplest events – and for me it will always produce these answers:

Nothing that we know of exists without a purpose and the overall purpose of our universe is to create, sustain and refine living things.

EIGHT PREDICTIONS

Scientific theories are supposed to be accompanied by predictions that bear out the ideas being proposed. As I've said, I'm no scientist, but the following are my "Einstein Wannabe" predictions. Although they are non-mathematical, very general, and in many cases unprovable for many years to come, they seem to fit naturally, reasonably and logically with the ideas I've proposed.

Prediction #1:
In the "near" future, scientists and astronomers will confirm virtually countless numbers of planets circling the trillions of suns in our universe.

Prediction #2:

Traditional religious beliefs will continue to wane as our knowledge of the Cosmos increases exponentially. This *may* lead to a period of violent conflict unless the word "God" can be re-defined.

Prediction #3:

Competition in all forms will slowly fade and eventually disappear from the lives of current and future generations.

Prediction #4:

The numbers of marriages and traditional family units will continue to drop substantially in this century and beyond.

Prediction #5:

Male and female roles in our society, and globally, will slowly coalesce, leading to an intersexual species.

Prediction #6 (Critically Important):
Developments in computer science and
digital technology will lead *to a merging
of human consciousness and
digital information.*

Prediction #7:
If we survive what may be the final stages
of our earthly evolution, we will undergo a new
form of change that will eliminate the cumbersome,
limiting aspects our animal ancestry (our bodies)
and leave us more suitably prepared to complete
the final phase(s) of our journey.

Prediction #8
In our final "human animal" phase, (which I
believe is "close" at hand), we will discover the
answers we seek:

**Why the immense, relentless effort
to create and refine life?**

**Who or *what* else is out there
related to this effort?**

Who or *What* is the designer?

D. Vincent, 2019

www.ingramcontent.com/pod-product-compliance
Lightning Source LLC
Chambersburg PA
CBHW030727180526
45157CB00008BA/3079